BALINESE
Flora & Fauna

JULIAN DAVISON
& BRUCE GRANQUIST

PERIPLUS

Publisher: Eric M. Oey
Text: Julian Davison
Illustrations: Bruce Granquist, Mubinas Hanafi, Nengah Enu
Production: Mary Chia, Violet Wong & Agnes Tan

Distributors
Indonesia:
PT Wira Mandala Pustaka
(Java Books–Indonesia)
Jalan Kelapa Gading Kirana
Blok A14 No. 17
Jakarta 14240

Asia Pacific:
Berkeley Books Pte. Ltd.
5 Little Road #08-01, Singapore 536983

United States:
Charles E. Tuttle Co., Inc.
RRI Box 231-5, North Clarendon
VT 05759-9700

Contents

Plate Tectonics

The Earth's surface is made up of gigantic rocky 'plates', which can be up to 100 kilometres thick. These plates are not 'fixed', but rest on more fluid layers beneath which allow them to move about, giving rise to the phenomenon known as continental drift.

Some 250 kilometres off the southern coast of Bali there is a very deep fault in the ocean floor known as the Java Trench. South of this lies the Indo-Australian plate which is moving northwards, crunching up against and sliding under the Sunda Plate on which Bali and the rest of Southeast Asia sits. As the Indo-Australian Plate dips under the Sunda Plate along the line of the Java Trench, periodic earthquakes occur, occasioned by sudden slippages between the two plates. At the same time, heat generated by the friction of these massive plates moving against each other causes pockets of molten rock to form under high pressure, and where these escape to the surface, they give rise to volcanoes.

Bali's Biogeography

Bali is a relatively 'young' island, having probably first emerged above the waves some 3 million years ago. It is one of the most volcanically active islands in the world and volcanoes have played an important role in shaping the natural history of Bali. Periodic eruptions spew forth great streams of molten balsatic lava which, when they have cooled, are broken down by natural erosion to form a fertile sediment in the valleys and plains. These rich soils supported the great swathe of forest that once covered most of the island and which can still be found in the west.

In historical times, much of this forest cover has been lost to agriculture, the natural fertility of the island allowing the cultivation of two rice crops a year to feed an ever-growing human population, which today stands at roughly 3 million.

Island Biogeography

Bali is a so-called continental island, that is to say, an island which sits on the continental shelf and is typically situated close to the mainland, in this case Java. Surrounded by shallow seas, it is periodically connected to the mainland at times of lowered sea levels which occur whenever there are ice ages. The latter circumstances allow an exchange of species—both plant and animal— with the mainland, with the result that the native flora and fauna of the island generally resembles that of the mainland, though the diversity of island species is usually less.

Bali is no exception to this rule, though the island does have at least one known endemic species, that is to say, a species which is unique to Bali and found nowhere else in the world. This is the famed Bali starling (*Lucospar rothschildi*) which was only discovered by Western science as recently as 1911. In the past, there was also a subspecies of tiger (*Panthera tigris balica*), but that has become extinct, a fate which also threatens the Bali starling.

Man and the Environment

The Balinese are an Austronesian people who share a common historical and cultural background with the other peoples of the region. It seems likely that the original homeland of the Austronesians was Taiwan, but some 6,000 years ago this ancestral population began to disperse, migrating southwards through the Philippines before turning eastwards into the Pacific and west towards the islands of Indonesia. This movement of people was not a mass exodus, but probably consisted of small groups of related kin, travelling by outrigger

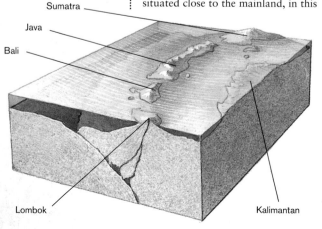

Sumatra
Java
Bali
Lombok
Kalimantan

5

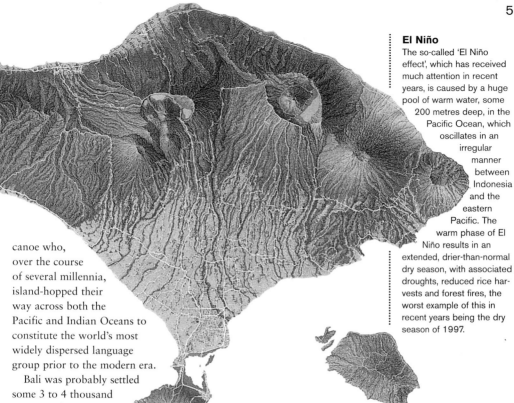

El Niño

The so-called 'El Niño effect', which has received much attention in recent years, is caused by a huge pool of warm water, some 200 metres deep, in the Pacific Ocean, which oscillates in an irregular manner between Indonesia and the eastern Pacific. The warm phase of El Niño results in an extended, drier-than-normal dry season, with associated droughts, reduced rice harvests and forest fires, the worst example of this in recent years being the dry season of 1997.

canoe who, over the course of several millennia, island-hopped their way across both the Pacific and Indian Oceans to constitute the world's most widely dispersed language group prior to the modern era.

Bali was probably settled some 3 to 4 thousand years ago by people who were farmers and herders, and who brought with them a bronze- and possibly iron-age technology. In the course of time, the human population has had a considerable impact on the natural history of the island, most notably in terms of deforestation and species extinction.

The most populated region has for centuries been the coastal plains and fertile valleys to the south of the central mountain range. It is here that the work of man is most evident in the form of the remarkable terraced rice fields which have been sculpted by hand from the natural contours of the land.

Rainfall and Climate Patterns

Climate plays a critical role in habitat formation and the dry southeast monsoon winds from Australia, which blow during the middle of the year, have an important influence on the type of vegetation found in the northeastern part of the island where lowland forest gives way to savanna grasslands.

Rainfall is also a crucial factor affecting vegetation types and growth patterns. On Bali, there are distinct seasonal variations, the wettest months of the year being between October and March.

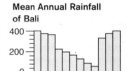

Mean Annual Rainfall of Bali

Months

Rainfall Distribution
A. < 1500 mm
B. 1500–2000 mm
C. 2000–3000 mm
D. 3000–4000 mm

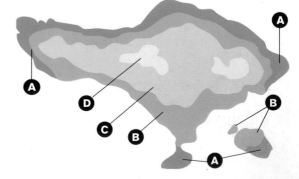

Volcanoes, Lakes and Caldera

Caldera Formation

Caldera are the result of a huge volcanic eruption, typically of an existing volcano which literally 'blows its top', pushing up an outer ring of mountains with a new cone at the centre. Often a lake will develop in the basin created by the outer ring of mountains.

The volcanic peaks of Bali are identified in myth and ritual as the home of the gods and like the Hindu deity Siwa, they are both the givers of life and the agents of death and destruction. On the one hand, their periodic eruption releases streams of molten lava and super-heated mudflows of cataclysmic proportions, while on the other, the erosion of balsatic lava deposits and their sedimentary redistribution in the valleys and plains is ultimately responsible for the rich soil that makes the island of Bali so naturally fecund.

The eastern end of Bali is dominated by the impressive symmetry of Gunung Agung (3,104 metres) whose summit is often ringed with clouds. In March 1963, Gunung Agung violently burst into life, showering its slopes with ash and debris which precipitated mudflows that killed thousands, left many more homeless, and laid waste to the surrounding countryside for miles around. This

was interpreted by some as evidence of divine displeasure at inappropriate ritual observances being celebrated at the temple of Besakih at the very moment of the eruption—Pura Besakih is situated on the slopes of Gunung Agung and is identified as the 'mother' temple for all Bali.

Lake Batur

Volcanic activity has been the principal cause of lake formation in Bali. Some are the craters of extinct volcanoes but Lake Batur, which is the largest, occupies a caldera—a volcano whose top has been blown off in a massive explosion, creating a plain, or in this case a lake, surrounded by a ring of mountains with a new cone at its centre.

Lake Batur lies at an altitude of 1,034 metres and covers some 16 square kilometres,

though the areas and depths of Bali's lakes vary with the seasons. At its deepest, Batur goes down to 88 metres and holds an estimated 815 million cubic metres of water. Ultimately though, the destiny of Batur, like all crater lakes, is to fill up with sediment though this process is likely to take thousands of years.

Batur Flora and Fauna

The arid slopes of Gunung Batur (1,717 metres), which rises at the centre of the Batur caldera, are hardened streams of balsatic lava and it will take several centuries for such deposits to be sufficiently weathered to produce soils that will support a closed forest. Even so, from the beginning, there are species of plant and animal which are quick to exploit the virgin terrain of a barely-cooled lava flow.

The most recent eruptions of Batur were in 1965 and 1974, but already mosses, grasses and ferns, typically the earliest colonisers of ash slopes, have long been in evidence. Their seeds (or spores in the case of mosses and ferns) are carried by the wind and settle in crevices in the jagged folds of the lava fields at the base of Gunung Batur where a little collected dirt will sustain them.

Lizards and skinks are among the first vertebrates to colonise the scorched landscape and may be seen warming themselves in the sun on exposed surfaces. As poikilotherms, or cold-blooded creatures, they like to take advantage of the sun's rays to warm their blood temperature which then enables them to move more rapidly in the pursuit of their prey and aids their digestion.

Freshwater Fish

There are no native fish species in Bali above an altitude of 500 metres, probably due to the low temperatures experienced at such heights. However, the non-indigenous giant gourami (*Osphronemus goramy*) has been introduced to Lake Batur and makes for good eating if dining in the vicinity of Kintamani.

Montane Plants and Animals

Tree ferns

Tree ferns, mainly *Cyathea* spp., are a common feature of montane forests, particularly lower montane forest. The delicate architectonics of their fronds and the geometric regularities of the diamond-shaped leaf bases which constitute their trunk, make them pleasing to the eye.

The tree fern is a very ancient plant type, being a latter-day representative of the great forests of tree ferns which once dominated the world's surface during the Carboniferous period some 300 million years ago. They are very slow growers, putting on 1 metre only every 15 years or so.

As one moves up from the plains and foothills to the central highlands, the landscape and vegetation changes noticeably. Heat is rapidly lost after sunset at high altitudes so that in the mountains, temperature ranges between daytime and night-time may differ by as much as 20 degrees centigrade. At the same time, the relative humidity of the air increases as the temperature falls, and at a certain point, condensation will occur causing drops of dew, and even clouds, to form. Much more rain is also likely to fall on the windward slopes of mountains, particularly if the prevailing air currents have passed over seas as is the case for Bali.

Montane Forest Morphology

The vegetation below 1,200 metres is very similar to that of lowland forests, but at higher altitudes, trees become increasingly stunted in height and epiphytes, which include many orchid species, become more common. At the highest altitudes, the trees look gnarled and stunted and there is an abundance of moss and lichens growing both on the ground and on the vegetation itself.

Eventually, these montane forests give way to a sub-alpine vegetation with even smaller trees and epiphytic lichens, but very few orchid species. Some alpine plant species are equipped with woolly hairs on their leaves which are thought to protect them against high temperatures during the day, low temperatures at night, and the intense ultraviolet radiation that is the result of the thinner atmosphere at high altitudes.

Roadside Monkeys

Monkeys are a common sight at the side of the road in the mountains of Bali. They are usually the long-tailed macaque (*Macaca fascicularis*), who typically travel in groups of 20–30 individuals, consisting of 2–4 adult males and 6–11 adult females, the rest being juvenile or adolescent members of the tribe.

The males are considerably heavier in build and larger in stature than the females, weighing between 5 and 7 kilograms. *Macaca fascicularis* are also found in coastal areas, especially mangrove forests, where they dine on crab, hence their alternative title of crab-eating macaque.

Casuarina

Casuarina junghuhniana is a conspicuous pioneer species on Balinese mountain sides. Like other pioneers, it only grows in extremely light conditions—*Casuarina equisetifolia* favours seashore locations—while its seeds will only germinate on bare soil and not within a forest environment.

Unusually for a pioneer, *Casuarina* is long-lived and can grow to considerable heights of up to 45 metres.

Although the *Casuarina* looks like a pine tree, with cone-like fruit and long feathery branches radiating from a single stem or trunk, its appearance is misleading. Whereas in the case of pine trees, the leaves are compressed into long needle-like cylinders, *Casuarina* 'needles' are actually slender twigs, the leaves themselves being reduced to minute, scale-like protuberances. In both instances, these modifications serve the same purpose, namely to reduce water loss through evaporation to an absolute minimum, which is a useful design feature in exposed situations with poor soils.

Like conifers, the *Casuarina* also drops its leaves in great profusion, and beneath its branches a dry tinder-bed of needles gathers, which readily ignites when occasional fires occur. The thick bark, however, is extremely fire-resistant and mature trees can withstand falls of hot ash during volcanic eruptions to sprout again when the cataclysm has passed. Often *Casuarina* will be the only tree for miles around, but it is also found in mixed montane forests.

CULTIVATED MOUNTAIN PLANTS

Coffee

The coffee plant is native to tropical Africa and was first introduced to Bali in the 1750s. There are two principal varieties—arabica and robusta—and both are grown in Bali. Arabica is more popular worldwide, but robusta is the preferred variety in Bali. The fragrant white flowers appear in April, but remain dormant for the next five or six months until the first rains of the southwest monsoon fall in September or October. The fruit, when it ripens, turns into red berries the size of small grapes, and is harvested at the beginning of the following year.

Cloves

Cloves are the unopened bud of *Syzgium aromaticum*, a member of the myrtle family, which are picked by hand and dried in the sun. Although a popular spice in Indian and even Western cuisines, cloves do not play an important part in Indonesian culinary traditions. Instead, their principal use is in the manufacture of clove cigarettes, or *kretek*, and as an analgesic agent—a little oil of cloves rubbed on the gums helps relieve toothache.

Cocoa

Although cocoa (*Theobroma cacao*) was first brought to Europe from the New World by Hernando Cortés as early as 1520, it was not introduced to Indonesia until the 1970s. Today, cocoa, or *coklat* as it is called locally, is quite commonly grown in the mountains of Bali, but it is not a popular beverage with the Balinese and is intended mainly for the export market.

Rivers and Streams

River Formation

None of the four main Balinese lakes—Batur, Tamblingan, Bratan and Buyan—has a surface outlet, but they do feed underground springs, which in turn give rise to rivers. Balinese rivers are for the most part fast-flowing, as they traverse from the mountains to the sea. They are, however, subject to seasonal fluctuations, the rivers of the west and northeast being dry for much of the year.

Water is a vital element in the natural history of Bali and one is seldom out of earshot of the sound of water in motion, whether it be the roar of a rushing mountain stream or the pleasant chuckle of the man-made watercourses and conduits which irrigate the rice fields.

The ultimate source of all freshwater is rainfall and this is distributed via three main pathways: overland, below the surface of the soil but above the water table, and in the groundwater. It eventually collects in a depression, be it a puddle, a pond or a lake, or alternatively flows, via streams and rivers, to the sea.

Mind you, plants take up quite a lot of the rain water from the soil which is subsequently returned to the atmosphere via pores in their leaves in the course of the natural cycle of transpiration. At the same time, a lot of surface water is also 'lost' to the atmosphere through evaporation, so that ultimately, only some 60–80 percent of the rain that falls on Bali actually ends up in the island's rivers and lakes.

River Fauna

The fact that none of Bali's rivers have long stretches of slow-moving water means that major

Kangkung

Water convulvulus, or *kangkung* (*Ipomoea aquatica*), is a popular local vegetable which grows in great profusion in less turbulent waters, including ponds and ditches, rice fields and, indeed, any other wet place. It may either be attached to dry land or else entirely free-floating: the buoyant stem sits on top of the water while the leaves are held a little above the surface. A relative of morning glory (*Ipomoea carnea*), it is easily recognised by its delicately tinted, almost white, flowers with purple centres, and its distinctive trefoil-shaped leaves. The latter, which resemble spinach when cooked, are usually served with a spicy prawn paste (*belacan*).

freshwater fish species such as carp, minnows and catfish are almost entirely absent from the island. Nor are there any native fish species found above an altitude of 500 metres on account of the cold, though at this altitude Bali's rivers and streams are teeming with invertebrate life—dragonfly and damselfly nymphs, water cockroaches, water boatmen, whirlygig beetles and countless other small aquatic beasts.

Whirlygig beetles, which belong to the Gyrinidae family of Coleoptera, are singular creatures in that each eye is divided in two, creating the impression that they possess four eyes. This arrangement in fact allows the whirlygig beetle to simultaneously scan both the surface of the water and its depths in search of a tasty morsel to prey upon.

Invertebrates are also prolific in the lower reaches of Balinese rivers which are generally poor in fish species. Indeed, the most common freshwater fish in Bali is probably the ubiquitous, but exotic, delta-tailed guppy, *Poecilia reticulata*, which was accidentally introduced from aquaria and now is found in almost every drop of water bigger than a puddle on the island.

Waterboatmen

Waterboatmen (Corixidae), have long oar-like hind legs equipped with hairs which they use to propel themselves across the surface of ponds and slower moving water. They prey upon mosquito larvae and augment their diet with plant matter. When they dive, short hairs growing around the abdomen trap a 'blanket' of air which they use for breathing under water. Waterboatmen also have wings and can readily take to the air to colonise new domains should a particular location not be to their liking.

Dragonflies

Dragonfly and damselfly nymphs are common in standing and flowing water where they dine voraciously upon tadpoles and small fish. Although they are slow movers, they possess a very effective hooked 'lip', or labium, which is equipped with claws and grapples for seizing their prey.

When a nymph reaches maturity, which can take up to a year, it crawls out of the water onto the stem or leaf of a plant, whereupon a small rent appears in the skin of its back from which the adult emerges.

Dragonflies spend most of their lives as nymphs, metamorphosing into the winged adult form as a prelude to mating. Death follows shortly after the completion of the reproductive cycle.

Lowland Rainforests

Bamboo

Bamboo is a giant type of grass with a worldwide distribution. A hundred species or more occur in Indonesia alone, and everywhere bamboo has been exploited by man for a great variety of purposes, ranging from building materials to water conduits, baskets, mats, fish traps, rice steamers, musical instruments and just about any other kind of container or implement imaginable. The hollow stems, or culms, are ideal for constructing lightweight structures which have both strength and flexibility. On the other hand, the young shoots make a delicious vegetable, although harvesting them destroys the plant.

Tropical lowland forests are among the world's richest and most diverse ecosystems, but everywhere they are under threat from man's activities. In Bali, the southern slopes of the central mountain range have long been given over to terraced rice fields, and since the late 19th century, rice crops for export have been planted on the drier northern slopes.

In this century, the principal cause of depredations to Bali's forest resources have been the commercial cultivation of coffee, clove and coconut via plantations. An increased demand for firewood to fuel brick, tile and quicklime kilns has also had a significant impact, while the volcanic eruptions of Gunung Agung and Batur have destroyed substantial areas of forest around their cones. Today, very little remains of the Balinese rainforest below 500 metres.

Forest Morphology

Rainforests are extremely complex habitats where the struggle for light plays a crucial role in determining the size and shape of the tree and plant species that grow there. In terms of its structure, or morphology, the Balinese rainforest is stratified into four or five distinct layers of vegetation. The upper storey, or forest canopy, consists of a continuous swathe of vegetation formed by the tops of mature tree species. Occasionally, where a tree has died and come crashing down, dragging some of its neighbours with it, natural clearings are formed, allowing light to reach the normally crepuscular forest floor. Here an entirely different range of species will flourish in the period it takes for the mature forest to regenerate. These include giant herbs like the banana and big-leafed tree species belonging to the genus *Macaranga*.

Above the canopy one sees, at intervals, trees of truly gigantic proportions, standing up to 80 metres tall, which erupt from the main forest canopy to stand head and shoulders above their competitors for light. Often there is a

lower storey just beneath the canopy comprising species which do well in the reduced light conditions. The forest floor, however, is relatively 'open', the dominant feature being slender seedlings vigorously sprouting from the leaf litter and decomposing organic matter which forms a thick carpet between the massive trunks of mature trees.

Emergent layer

The need to reach the canopy as quickly and efficiently as possible means that the majority of rainforest tree species have very tall, straight trunks, which soar vertically for 50 to 60 metres, sometimes more, before branching into a crown of foliage. This distinctive tree profile is a characteristic feature of the rainforest and is directly attributable to the competition for light.

Unlike the lowland forests of Sumatra and Kalimantan, the surviving rainforests of Bali are not characterised by the presence of a dominant family or species of tree.

Nor is there a consistent 'mix' of forest composition in terms of typical species found within a given area.

The Forest Floor

Death and decay are the means by which organic materials and nutrients are recycled back into the system. A lot of this goes on at ground level where beetles and fungi are engaged around the clock breaking down dead leaves, fallen tree trunks and other plant matter into a nutrient-rich humus.

The Basidiomycetes, which include the genus *Amanita*, are the most prominent class of fungi found in the rainforest, being easily recognised by their large, spore-producing fruiting bodies which come in a wide range of colours, shapes and sizes. Basidiomycetes fungi play an integral part in the decomposition of wood and typically occur in clusters on fallen branches and dead tree trunks.

Amanita spp.

Water Cycle

Rainforests occur in areas which are wet almost all year round, and where the dry season, if any, is limited to no more than a couple of months. In Bali, they are found on the lower southern and western slopes of the central mountain range, which provides a natural shield against the drier seasonal weather in the north and east parts of the island.

Water plays a vital role in sustaining the rainforest, being an essential ingredient in the generation of plant energy through photosynthesis. Consequently, the leaves of many rainforest tree species are specially adapted to ensure that the maximum amount of water precipitated during rainfall reaches the root system. This adaptation takes the form of a 'drip tip' which effectively drains the leaf of collected rain water and directs this precipitate to the ground at the base of the tree to be subsequently absorbed by the roots.

Drip tip

Forest Fauna

Bali is home to quite a wide range of forest animals, though this may not always be immediately apparent to the casual observer, given the nocturnal habits of many forest species and the ample cover provided by dense tropical vegetation. In the past, the most celebrated of these was the famous subspecies of tiger, *Panthera tigris balica*, but unfortunately this magnificent creature, like its Javanese cousin *Panthera tigris sondaica*, has become extinct. Diminishing areas of natural habitat mean that a number of other forest species are similarly under threat, particularly in the case of larger creatures which generally need more area to survive as self-sustaining breeding populations.

The biggest animal on the island is the *banteng* (*Bos javanicus*), which is a species of wild cattle found in several parts of Southeast Asia. Other notable mammals include the Javan *lutung* (*Semnopithecus auratus*), a type of monkey endemic to Java and Bali, and the Javan fer-

ret-badger (*Melogale orientalis*), which is also endemic to the two islands. Of the three species of deer, the most remarkable is the minuscule mouse deer (*Tragulus javanicus*) which weighs less than two and a half kilograms and seldom measures more than half a metre from its nose to the tip of its tail.

Bali boasts a large number of snake species, of which about 30 percent are potentially dangerous to man. These include the venomous triangular-headed pit vipers, various species of cobra, krait and coral snakes, a number of sea serpents, and the non-venomous but constricting, reticulated python (*Python reticulatus*). The latter is one of the largest snakes in the world. Lacking venom, pythons kill their victims by wrapping their coils around the unwary and then crushing the breath out of them. It seems that it is brain death through oxygen deprivation, rather than actual suffocation, which causes death. Pythons are mainly nocturnal and can be found in most habitats, including gardens, but only the larger specimens pose a threat to humans.

Flying Foxes

Flying foxes are actually large fruit bats, the most common species in Bali being *Pteropus vampyrus*, a rather unfair taxonomic epithet for a creature of such inoffensive dietary habits.

Cattle

On Bali, the largest extant representative of the island's indigenous fauna is the *banteng* (*Bos javanicus*), a species of wild cattle which can reach up to 1.8 metres at the shoulder and weighs some 900 kilograms. Probably less than 20 of these beautiful beasts still roam free in the Bali Barat National Park, and it is possible that the purity of the herd has been contaminated by mating with the smaller, domesticated Balinese *sapi*, which is descended from the wild *banteng*.

Forest Bird Species

Forest birds are well represented in Bali and it is quite easy to spot a fair number of indigenous species from the roadside if one takes a route through the central highlands, particularly in the vicinity of the three crater lakes of Bratan, Buyan and Tamblingan. The Bali Barat National Park is also a refuge for many native forest species.

Greater Racket-tailed Drongo

The greater racket-tailed drongo (*Dicrurus paradiseus*) is fairly common in areas of Bali where there are remnants of lowland forest. This spectacular bird is easily recognised by its distinctive forked tail which trails some 30–40 centimetres behind. Tweedie, in his *Common Birds of the Malay Peninsula*, comments that "When the bird is in flight, the shafts [of the tail feathers] are barely visible and one receives the impression of a black bird closely pursued by two large bees."

Greater racket-tailed drongo
(*Dicrurus paradiseus*)

Bali Starling

The most famous bird in Bali, and one that is extremely unlikely to be seen from the roadside, is the endemic Bali starling (*Leucopsar rothschildi*). First discovered by European ornithologists in 1911, its distribution, even at that time, seems to have been limited to the western tip of the island. Today, it is a strictly protected bird and there are severe penalties for those who try to capture them for the caged bird market. Despite these measures, at the last count (1993), there were only 34 individuals left in the wild and it is perhaps fortunate, though somewhat ironic, that there are now probably several thousand in captivity—there are plans for a controlled breeding programme which will allow the release of their offspring back into the wild.

Magpie Robin

The magpie robin (*Copsychus saularis*) is a common denizen of Balinese gardens and hotel lawns, where it is frequently seen, particularly in the early morning, strutting around in its dapper pied plumage, with tail fanned out, plucking earthworms from the moist topsoil. Its call is a loud clear whistle, which is variously phrased and issues forth in a continuous chortle, proclaiming its territorial rights to any would be challengers. Disputes with other males are characterised by a harsh acrimonious chatter.

Female magpie robins are slightly less distinguished than their mates. Whereas the males sport a blue-black plumage, this is replaced by dull grey feathers in the females.

Bali starling
(*Leucopsar rothschildi*)

Magpie robin
(*Copsychus saularis*)

Deciduous Forests and Savanna Grasslands

Opuntia nigrans
The 'spines' of the prickly pear cactus (*Opuntia nigrans*) are actually its leaves, which have become modified to reduce water loss. It is not a native of Bali, but is ideally suited to the drier northeastern corner of the island where it flourishes in great profusion.

Deciduous forests, composed of trees which shed their leaves periodically, occur in those parts of Bali where there is a prolonged dry season of several months' duration. This is mainly on the northern side of the central mountain range and the terrain becomes increasingly drier and subject to drought as one heads towards the northeastern corner of the island.

The trees which grow in these dry areas have necessarily developed deep root systems and are much smaller in stature than in rainforests—few reach above 25 metres. At ground level, annual plant species with underground tubers are common, while many plants have leaves and stems covered in spines or thorns to protect them from the predations of grazing herbivores. The coastal savanna forests of northwest Bali are characterised by the flat-topped crowns of

Acacia trees, interspersed with *lontar* palms (*Borassus flabellifer*). The northeastern part of the island, however, is much drier because the rainfall percolates rapidly through the ash-cinder soil laid down by successive eruptions of Gunung Agung. Here, the deciduous forests have given way to a savanna vegetation consisting of extensive grasslands with a discontinuous tree layer dominated by the ubiquitous *lontar* palm.

Common Deciduous Species
Leguminous species, that is, pod-bearing trees and plants such *Albizia*, *Acacia* and *Cassia*, are common in the arid northeast. Many of these species are dormant during the driest months, suspending their growth until the arrival of the seasonal rains. When the latter break, there is a sudden burst of floral activity which results in the overnight appearance of a mass of flowers and foliage that dress the forest from floor to canopy.

Kapok Trees

The *kapok* tree, with its distinctive horizontal branches, is another common feature of Bali's drier regions. Between October and November, the branches hang heavy with pods, each bearing tufts of white, fluffy fibres, like cotton wool, which are used to stuff mattresses and pillows.

Two Species of Palm

The deciduous forests of Bali are dominated by two rather similar-looking palm species—the *lontar* palm (*Borassus flabellifer*) and *Corypha utan*. The latter is the larger of the two species, with leaves that can grow up to nearly 2 metres in diameter. The *lontar* or palmyra palm is the more common of the two species, especially in the drier, northern parts of the island where it is known as *punyan ental*. It is distinguished by its enormous fruit which can grow up to 18 centimetres in diameter.

Palm Wine and Sacred Manuscripts

In the northeast of Bali, the sap of the *lontar* palm is used to make an alcoholic beverage known as *tuak*. A cut is made in the growing tip of the spadix (the stalk bearing the flowers) and the 'juice' drawn off. A dash of last night's *tuak* is then added to encourage fermentation, with different grades of alcohol being produced according to the length of the time allowed. Elsewhere in Bali, *tuak* is made from the sap of the coconut palm.

The immature leaf blades of the *lontar* palm were traditionally used as a writing material on which sacred Hindu and Buddhist texts were inscribed—the Indonesian word *lontar* is derived from the Javanese term for 'word', *ron*, combined with *tala*, meaning 'leaves used for writing'. The dried fronds also serve as ready-rolled cigarette 'papers'.

Lontar palm and fruit

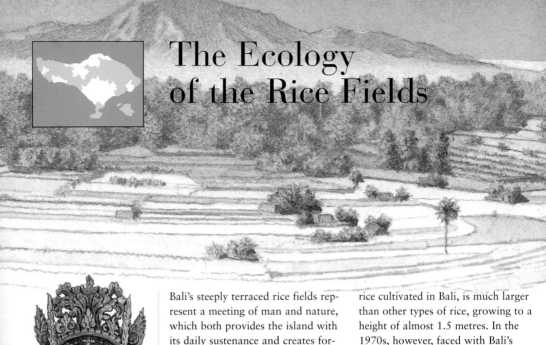

The Ecology of the Rice Fields

Bali's steeply terraced rice fields represent a meeting of man and nature, which both provides the island with its daily sustenance and creates formal patterns of astonishing beauty and variety. These irrigated terraces allow two crops of rice to be planted each year, with water from the mountains being directed to each individual rice field by an intricate network of channels and aqueducts.

Subak Councils

These man-made waterways are maintained and regulated by local cooperative organisations called *subak*. Each mini-watershed has its own *subak* council, made up of neighbouring farmers, and they are responsible for the equable distribution of water to all the irrigated rice fields within their purview.

Each *subak* council has its own temple, situated in the middle of rice fields, where major ceremonies of the rice cycle are held. The temple of Ulun Danau, on Lake Beratan, is identified as the 'mother' temple of all the *subak* systems on the island.

'New' versus Traditional Strains of Rice

Padi Bali, the traditional variety of rice cultivated in Bali, is much larger than other types of rice, growing to a height of almost 1.5 metres. In the 1970s, however, faced with Bali's rapidly rising population, local farmers were encouraged to plant 'new' strains of rice, first developed in the Philippines, which were high-yielding, faster-growing and more resistant to disease and insects. The success of these new rice strains have reversed Bali's agricultural economy from an annual shortfall in the 1970s to a surplus production that allows the net export of several tens of thousands of tons of rice each year.

Rice and Ritual

Ritual plays a key role in the cultivation of rice in Bali, each stage in the agricultural cycle being underpinned by ritual prescriptions and prohibitions. As in many parts of Southeast Asia, there is a general association between the fertility of rice and that of women—the ripening panicles of rice are said to be 'pregnant' (*beling*) and the most important deity associated with the cultivation of rice is the goddess Dewi Sri. Images of the latter, constructed from plaited rice stalks, are everywhere in the rice fields at harvest time and can also be

Sacred Rice

Rice farming is central to the Balinese way of life. It is something sacred; it is life itself. According to Hindu mythology, rice was created by the god Vishnu and is man's divine patrimony which must be cultivated according to instructions handed down from Indra. A key figure in this scheme of things is the rice goddess Dewi Sri (above), wife of Vishnu and paragon of all that is good and beautiful.

seen attached to rice barns.

The most lavish ceremony in the rice cycle is the *ngusaba nini* ritual, which is held at the *subak* temple, either just before or after the rice harvest. It is in effect, a thanksgiving festival dedicated to the rice goddess Dewi Sri and involves the preparation of sumptuous offerings of food which are presented to the deity in recognition of her beneficence.

The Rice Cycle

The rice cycle begins with the breaking up of the dry ground and leftover stubble from the previous season. The field is then hoed and filled with 5 to 10 centimetres of water before being ploughed.

Meanwhile, the rice seed is broadcast by hand in a specially prepared seed bed or nursery. After germination, the young rice plants are allowed to grow for three to four weeks before being transplanted to the rice field. When re-planting the seedlings, the space left between individual plants is critical. Closely-spaced plants restrict the growth of weeds, but if the interval is too small, then the growth of rice plants themselves may be affected.

Some plant species other than rice—for example, *kangkung* (*Ipomoea aquatica*)—are actively encouraged to grow in the rice field since they may be eaten as vegetables. Fish are also a good thing—they feed on algae and other weeds, converting these into fertiliser which further enhances the growth of rice.

The Rice Harvest

The rice crop takes roughly 120 days to reach maturity (*padi Bali* requires a month or so longer) and is harvested by hand. The newer strains of rice are simply threshed on the spot but *padi Bali* is tied into beautiful round bales before being taken back to the rice barn where it is ceremoniously received.

Back in the rice fields, the stubble and leftover straw is torched, returning valuable minerals to the soil for the next season's crop.

Ancient Cultivars

Rice (*Oryza sativa*) is one of the few crops which thrives in water-saturated soils and can even be submerged for part or all of its growth cycle. Its cultivation in Asia is of considerable antiquity and today there are literally thousands of different varieties, reflecting millennia of selection on the part of countless generations of rice farmers.

Rice Field Fauna

The sublime spectacle of Balinese rice fields and terraces usually encourages reflections of a tranquil and meditative nature, but amongst those luscious, viridian-green rice stalks, and beneath the still waters of the flooded padi field rages a ferocious battle for survival. Monstrous predators gorge themselves on hapless prey, while voracious herbivores feast on the maturing rice plants with the rapacity of runaway lawnmowers. No Garden of Eden this!

Rice Pests

Insects are a considerable menace to the success of the rice crop and they come in a variety of forms and dietary preferences. There are stem borers, leaf-folders, leafhoppers, planthoppers and any number of bugs which simply enjoy eating rice as much as the Balinese. The brown plant-hopper (*Nilaparvarta lugens*) is an especially noisome beast which not only damages the rice plant by sucking sap from the vascular system of the leaves, but also, by this means, introduces viral infections directly to the plant. The seriousness of the damage inflicted by brown planthoppers is exacerbated by the fact that they are able to reproduce themselves in vast numbers and with great alacrity. Females can lay between 100–300 eggs in a three week laying period, with only four weeks between successive generations, while males are ready to mate just one day after hatching and go on to survive for almost a month subsequently.

What is more, the eggs of the brown planthopper are laid deep in the leaf sheath of the rice plant which means that pesticides are unlikely to reach them.

Actually pesticides are part of the problem, rather than the solution, in that they do not discriminate between 'good' bugs and 'baddies' so that their widespread use in recent years has decimated the natural predators of planthoppers. In the past, when rice was grown under more 'natural' conditions, planthoppers were subject to the attentions of over one hundred different kinds of predators, parasites and diseases, which kept their numbers under control.

Spiders

But planthoppers don't have it all their own way, being one step down

Green leafhopper
(*Neophotettex virescens*)

be recognised from the fork-shaped marking on its back. These Arachnids do not spin webs but lurk instead at the base of rice parts ever-ready to jump upon and devour their unsuspecting prey.

Spiderlings feed upon plant-hopper and leafhopper nymphs, while adult wolf spiders indicate a preference for stem borer moths. Each individual wolf spider can dispatch between five to 15 victims a day and in this respect, they are an extremely effective natural agent of pest control.

Birds

Birds can also play a useful role in controlling pestilence in the rice fields—cattle egrets (*Bulbulcus ibis*) and the Javan pond heron (*Ardeola speciosa*) are two easily spotted species which commonly occur in Balinese rice fields. Egrets feed on grasshoppers, crickets and spiders, while the Javan pond heron prefers dragonflies and water beetle larvae, or spiders and molecrickets. Molecrickets (*Gryllotalpa* spp.) are a particular nuisance because they live among the roots of the rice plant, feeding on the lower stem and loosening the surrounding soil, causing the plant to wilt.

Frogs

Frogs find rice fields much to their liking and a variety of species play their part in containing rice pests by dining on grasshoppers, crickets, beetles, bugs, and butterfly and moth larvae.

in the food chain to spiders. Spiders are the most aggressive predators on planthoppers and other small creatures which afflict the rice crop, and their presence is a welcome one in the rice fields.

The aptly-named wolf spider (*Pardosa pseudannulata*) is particularly rapacious and can easily

Ducks

Ducks enjoy a diet of fresh-water snails and Balinese farmers are mindful of this, escorting their ducks each day to their rice fields and allowing them to get on with the job.

To prevent the ducks from straying too far in the absence of the farmer, a stick with a small white flag topped by a bunch of white feathers is planted in the mud. Balinese ducks are 'fixated' by this marker and will follow wheresoever it leads them.

At dusk, they dutifully gather round the base of the flag, awaiting the signal to march home again. All that the duck-keeper has to do is simply to uproot the staff and walk back through the rice fields to his family compound, assured that his flock of ducks will follow closely at his heels.

Gardens and Compounds

The Balinese are keen horticulturists as well as rice farmers. In addition to growing a variety of fruit and vegetables, they are also fond of colourful flowering plants and shrubs which enhance the beauty of their homes and villages. In this last respect, they have been assisted by the government's *tamanisasi* policy which is aimed at turning Bali into a garden isle. This has been implemented on a communal basis with coordinated planting schemes outside people's homes and by the side of the road.

In terms of variety, relatively few plant species are found in such settings compared to the profusion of species occurring in the wild. Furthermore, many of these ornamental plant species are 'exotics', introduced to the island from elsewhere for their decorative or shade-creating qualities. Exotic tree species include the magnificent rain tree (*Albizia saman*) from tropical America and the flamboyant or flame tree (*Delonix regia*) from Madagascar. The former gets its name from the fact that with the approach of rain, it collapses its leaves, thereby ensuring that maximum rainfall reaches its root system.

Hibiscus
The flowering shrub hibiscus (*Hibiscus rosa-sinensis*) originally came from China, but today is found everywhere in the tropics and is grown in almost every Balinese compound or garden. Its flowers,

which are produced each day in great profusion, are used in offerings and to decorate temples, statues and other religious structures. They are also often tucked behind the ear as a personal adornment. A wide range of colours are available, but the most common is the bright red variety.

Frangipani
The frangipani (*Plumeria* spp.) is a native of the Americas and was first introduced to Southeast Asia by the Spaniards via the Philippines. The most widespread species of frangipani in Bali, *Plumeria obtusa*, is easily recognisable by its waxy white flowers with yellow centres, and its long leathery leaves with rounded ends. *Plumeria acuminata*, with its pinkish blossom, mixed with yellow and white, is also commonly planted, particularly in temple precincts and Muslim cemeteries.

Bougainvillea
Bougainvillea (*Bougainvillea* spp.) is a native of Brazil and is named after the French navigator Louis Antoine de Bougainville.

The flowers, which grow in groups of three, are actually quite inconspicuous, being thin, tubular structures with a slightly flared mouth. Rather, it is the often brilliantly coloured

Heliconia
(*Heliconia rostrata*)
Heliconia flowers are relatively insignificant, depending on the species; rather it is the brightly coloured bracts arrayed along an extended inflorescence which provide such a spectacular floral display.

bracts—the leaf-like structures that surround the flowers—which provide the floral display. The latter have a papery quality which gives the bougainvillea its Balinese name of *kertas* (literally, 'paper'). Some species are always in flower; others only bloom in response to dry seasons. The most spectacular example is the aptly-named *B. magnifica*, which has large magenta-coloured 'flowers' and blooms all year round.

side tree, partly for its shade dispensing properties, but also for its brilliant red flowers. The latter appear towards the end of the dry season, first blooming in mid-September, with the greatest profusion in November. The flowers are eventually replaced by seeds contained in dark brown pods.

Butterfly
(*Ornithoptera paradisa*)

Flamboyant

The flamboyant or flame tree (*Delonix regia*) is a native of Madagascar but was introduced to Bali in the 19th century where, like the rain tree, it is planted as a road-

Flamboyant or flame tree
(*Delonix regia*)

CULTIVATED FRUIT

Mangosteen

Mangosteens (*Garcinia mangostana*) are one of the most delicious fruit, though not well known outside the tropics as they do not travel well. Slow growing, they are difficult to cultivate commercially, taking up to 15 years before bearing fruit with only a few ripening at a time. Famous for their 'cooling' properties in terms of local theories of bodily health and vigour, mangosteen, in many parts of Southeast Asia, are also regarded as an aphrodisiac.

Salak

The fruit of the *salak* palm (*Salacca zalacca*) is covered with a thin, but tough, outer skin consisting of hundreds of tiny triangular scales. This external covering closely resembles the skin of a reptile, both in pattern and texture, and *salak* are sometimes referred to as 'snake fruit' in English.

Salak grow in clusters and each fruit is segmented into three or more lobes, each containing a hard shiny seed. The flesh is crisp, like that of a radish, and the taste is singular—not sweet, but refreshing. The architectonics of the *salak* palm reveal a very ancient ancestry, being virtually stemless with a primitive rosette pattern of growth. The *salak* palm is also distinguished by a vicious display of thorns, some up to 4 or 5 centimetres in length, which can make harvesting of its fruit a painful business.

Mango

The succulent mango (*Mangifera indicus*) is native to the rainforests of Southeast Asia, and despite its global distribution today, the best varieties are still to be found in the region, reflecting literally thousands of years of cultivation and selective propagation by man. The tree is very attractive, with long, spear-shaped leaves which are a delicate shade of pink or even violet when newly formed, depending on the species. Tiny flowers appear in September and the fruit, which hang down on slender stalks, ripen by November.

There are a great number of different varieties of mango which come in a range of shapes and sizes. Not all are as good as the other, but the best are incomparable, with the *poh golek*, or 'round mango', generally deemed to be the tastiest according to local palates.

Animals and Man

Animals are very much a part of everyday life in Bali. Some, like the water buffalo, are domesticated species; others simply occupy the same space as man and are either tolerated or exterminated. Most unfortunate of creatures is the Balinese dog—a wretched and ulcerous brute, maltreated and unloved, yet a ubiquitous and often alarming feature of every Balinese village. Not every animal, however has such a miserable existence, and there are some which are especially privileged—the cow and the fighting cock, for example, are the subject of lavish care and attention.

Cage Birds

The keeping of birds in cages is a very ancient tradition in Indonesia. They may be kept for their song, their aptitude for mimicry, their soothing cooing, or just simply because they are beautiful to look at.

Favourite song birds include the Asian pied starling (*Sturnus contra*), black-naped orioles (*Oriolus chinensis*) and the white-rumped shama (*Copsychus malabricus*), while cockatoos and hill mynahs (*Gracula religiosa*) make good mimics. Other species, like the spotted dove (*Streptopelia chinensis*) (above), fill the air with mellifluous cooing sounds, while the Java sparrow (*Padda oryzivora*) and a variety of munias (*Lonchura* spp.) are admired because of their attractive appearance.

Pot-bellied Pigs

The low-slung Balinese pig, with its sagging back and pot belly virtually dragging along the ground, was once a characteristic sight in Bali, but is far less common today through the introduction of larger British breeds.

Balinese porkers are generally kept in pens inside the family compound, though when small they may be simply tethered to a secure point. They are looked after by the women of the household who feed them a diet of rice hulls mixed with water, supplemented with banana stems.

Bali Cattle

The graceful, faun-like native cattle of Bali are a domesticated, scaled-down version of the wild *banteng* (*Bos javanicus*). They are husbanded by men who can often be seen in the evenings lovingly sluicing down their beasts with water from an irrigation canal. Although Hinduism is a fundamental part of Balinese religion, many people will eat beef, though not if they are priests or otherwise engaged in religious duties. Those who do eat beef may, however, shampoo themselves afterwards, the head being considered the most sacred part of the body. Milk is not a problem one way or another.

Fighting Cocks

Betting on fighting cocks, and indeed all other forms of gambling, are illegal in Indonesia, but cockfighting (*tajen*) has very deep and ancient roots in Balinese culture, and contests continue to be held regularly on the island, albeit discreetly out of sight of the authorities.

Fighting cocks, like all domesticated poultry, are descended from the wild jungle fowl (*Gallus gallus*) which is native to the region. They are greatly esteemed by their owners

who feed them on a special diet of top quality maize.

Equipped with razor-sharp steel spurs some 10 centimetres long, the fighting cock is a lethal adversary in the ring and often both contestants are mortally wounded, in which case the victor is he who expires last. Intense betting surrounds the fight. The stakes are high and bets must be settled immediately—a man can literally lose his shirt.

The cockfight, because it involves the spilling of blood, has a symbolic aspect to it, being seen as a propitiation to malevolent spirit influences (*bhuta* or *kala*). Because of this ritual element, the Indonesian gov-

ernment allows cock fights to be held on ceremonial occasions provided no betting takes place.

Rats

The most successful animal species in Bali are rats, the majority of whom lead quiet, unassuming lives in the forests and other wild areas. The native house or roof rat (*Rattus rattus*) and the rice field rat (*Rattus argentiventer*) do, however, constitute a serious problem.

Traditionally, at the end of a harvest, the Balinese will dig up all the rat nests they can find in the rice fields and club as many rats to death as possible. Two of their number, however, are allowed to survive and are released back into the rice field in the hope of gaining forgiveness from the spirits of their less fortunate brethren whose bodies are then cremated.

Despite these measures, it was felt, not so long ago, that for every rat bludgeoned to death, four more would spring up on the spot where it died because the Chief Mouse was piqued at not having received sufficient sacrifices. A special ceremony was therefore held at Besakih, the 'mother temple' of all Bali, to appease this disaffected rodent deity. Afterwards, rat poison was distributed to rice farmers by government agencies, but the results were only a qualified success.

Cockroaches

One creature the visitor to Bali is unlikely to miss is the cockroach, which will typically be encountered lurking in the corner of a damp bathroom, waving its antennae in a slightly menacing manner. The most common species on the island are *Periplaneta australasiae* and *P. americana*, which despite their taxonomic designations were probably first transported to the region from Africa.

Cockroaches are very hard to eradicate because they are extremely sensitive to changes in their environment. Finely-tuned sensory organs in their antennae are able to detect minute changes in air pressure, temperature and moisture, while equally sensitive receptors in their legs can pick up the slightest vibration. Although cockroaches can fly, their preferred mode of locomotion is to scuttle, which they are able to do with great facility and speed.

Tokay Gecko

The tokay gecko (*Gekko gecko*) is a familiar sight and sound in Bali with its instantly recognisable call—a slightly disconsolate 'uh-uh'. Despite its relatively large mass—they can grow up to 35 centimetres in length—the tokay gecko can traverse a ceiling as sure-footedly as any of its lesser brethren, even when some of its limbs are missing. Tokay geckos are further distinguished by their unblinking stare (their eyelids cannot move) and their ability to wilfully detach their tail to provide a distraction to pursuing predators or as a last-ditch means of escape if seized by this appendage.

The Coastline and Intertidal Zone

The coconut palm (*Cocos nucifera*) is the quintessential feature of any tropical shoreline—a symbol of romance and earthly paradise to some, a natural resource of great nutritional and utilitarian value to others. Its wide distribution may be partly accounted for by the fact that the fruit of the coconut palm—the coconut itself—comes in a thick, buoyant husk which can keep the seed fresh for several months, even at sea and in high temperatures. Consequently, it can be transported for thousands of miles by ocean currents and still remain viable. Man, too, has probably played an important role in the distribution of the coconut palm: its good storage qualities make it an ideal source of food and water on long ocean crossings.

As a source of food, the growing tip, or 'heart', of the coconut may be eaten lightly boiled as a vegetable, or else raw in a salad, while the fermented sap, drawn from the spadix or flower stem, produces a toddy called *tuak*, which can also be distilled into a liquor (*arak*). Alternatively, a syrup or sugar may be obtained if the water content is removed by evaporation

The flesh of the coconut is good to eat but can also be turned into an oil which may be used for cooking or else applied to the skin and hair. Dried as copra, the meat is used in the manufacture of soap and margarine, while the outer husk produces a natural fibre called coir, which is used to make ropes and door mats, or as a stuffing for mattresses. The ripened nut also contains a clear liquid which provides a refreshing drink or cocktail base. Lastly, the leaves of the coconut tree may be woven into mats or roofing thatch.

Birdlife

A variety of birds can be seen along the shoreline. They include waders, grubbing in the sand and mud for their daily fare of small crabs, fish larvae, worms and bivalves; coastal raptors, such as the white-bellied sea eagle (*Haliaeetus leucogaster*), the

alidris alba

Brahminy kite (*Haliastur indus*) and ospreys (*Pandion haliaetus*); and various seabirds, such as boobies, frigatebirds, tropicbirds, terns and noddies.

The latter can spend many months at sea without returning to land. When they do so to mate, they tend to nest in large colonies on rocky cliffs or small islands where there are few predators: Ulu Watu and Nusa Penida are popular Balinese nesting sites for these ocean-going species.

Mangroves

Mangrove forests are an important feature of the Balinese coastline, providing nursery grounds for young fish, shrimp and other marine crea-

tures and helping to prevent coastal erosion by absorbing onshore waves.

The soil in mangrove forests is waterlogged and without oxygen, so native plant species have evolved elaborate root systems which allow them to 'breathe'. The stilt roots of *Rhizophora* species are one solution to this problem, while *Bruguiera* species have developed roots which loop in and out of the soil forming aerated 'knees' above ground level.

Other local specialisations include the self-planting fruit of *Rhizophora* species. Unusually, the seed actually germinates whilst still attached to the parent tree. A long torpedo-like root soon develops and when the seedling, with its first leaves emerging, drops from the branch, this structure readily penetrates the soft mud below.

Rhizophora spp; mature tree, fruit and flower.

Mangrove Crabs

Mangrove forests are home to large numbers of crabs who survive by eating the fallen leaves of mangrove tree species. It is estimated that up to 90 percent of fallen leaves in mangrove forests are either eaten on the spot by crabs or dragged into their holes within three weeks of dropping to the ground. Digested by the crabs, they enter back into the organic cycle as excreted detritus which is further enriched by fungi, bacteria and algae growing on or in it.

This detritus is an important item in the diet of a great many small creatures, including zooplankton, prawns, other crabs and small fish, which in turn feature on the menus of larger animals with carnivorous appetites—mainly species of fish.

Coral Reefs

Coral reefs are extremely complex communities built around the compacted and cemented skeletal remains of marine creatures and the limestone secretions of sedentary organisms, some of which are the coral itself. The latter requires lots of light and warm temperatures, which is why coral reefs are only found in tropical waters and usually not below depths of 100–120 metres.

Polyps and Plankton

The coral animal starts life as a small planktonic larva which, when it comes to rest on a suitable substrate, metamorphoses into a polyp—a tiny animal rather like a sea-anemone. This polyp begins dividing asexually, producing countless genetically identical individuals, or clones of itself, which over time grow into large conglomerations. Coral polyps secrete a limestone external skeleton. As they die, new polyps form on their skeletal remains, resulting in the steady expansion of the coral reef.

By way of sustenance, coral polyps feed off plankton and other tiny sea creatures which they ensnare in their hydra-like tentacles. Small photosynthetic organisms—golden-brown algae—live within the 'skin' of the coral polyp, absorbing waste products released by the polyp while using the sun's energy to convert phosphates and nitrates into protein, and carbon dioxide into oxygen which is used by the coral polyp for its respiration.

Coral polyps reach sexual maturity after about three years, releasing their gametes simultaneously into the surrounding seas in order to ensure the maximum opportunity for eggs to meet with sperm. This mass reproduction typically occurs at night, usually a few days after a full moon.

Sea Anemones

Sea anemones are very similar creatures to coral, but unlike the latter, do not possess a calcareous external skeleton. A constant circulation of water through the central cavity of the anemone brings them a host of minute marine organisms to dine upon, although like corals, sea anemones receive most of their nutrition from photosynthetic algae living in their tissues.

To assist them to stun their larger victims, sea anemones are equipped with a vicious armoury of stinging cells which discharge a barbed, poison-tipped thread into anything that comes into contact with them. Not that this deters the passing sea slug who feasts on sea anemones without undue discomfort. Indeed, when an anemone is eaten by a sea slug, its stinging cells are not digested but

Reef Formation

Fringe reefs and barrier reefs represent successive stages in the development of a reef. The former tend to be relatively young and are associated with rock coastlines. They typically occur quite close to the shore and extend themselves seawards, with a shallow 'reef flat' forming between the crest of the reef and the land. Coral growth on the reef flat is poorer because of reduced water circulation, sedimentation and freshwater runoff from the land. This means that as the reef grows outwards towards the open sea, the innermost corals cannot keep pace and a lagoon develops between the outer margin of the reef and the shoreline. At this point, the fringe reef has become a barrier reef.

Snappers

Snappers (*Lutjanus* spp.) are good to eat and are among the region's most important commercial fish. Most species favour inshore waters and reefs, but some are found at sea and at considerable depths.

Sea fan

Sea anemone

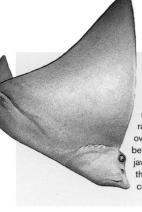

Eagle Ray

The flattened body and wave-like motion of the spotted eagle ray (*Aetobatus narinari*) are designed for a life on the sea bed, and they are typically found wafting across seagrass meadows and sandy areas of coral reefs. Eagle rays can grow to more than 2 metres in length and weigh over 200 kilograms. The long slender tail is armed with between two and six venomous barbs, while their powerful jaws are equipped with large, plate-like teeth which enable them to crush the chitinous armour of the molluscs and crustacea which constitute their daily diet.

Anemonefish

Clown fish, and other species of anemonefish (*Amphiprion* spp.), have struck up a unique relationship with sea anemones which enables them to live amongst the anemone's tentacles without being harmed by their stinging cells.

The fish acquire this immunity by secreting an enveloping mucus which absorbs a substance from the anemone, carrying the latter's chemical signature. When the anemonefish subsequently comes into contact with the anemone, the stinging cells are prevented from firing by the same chemical inhibitor mechanism that prevents one tentacle from stinging another.

are passed through the sea slug's body to be incorporated in the latter's tentacle-like projections (*cerata*) as part of the sea slug's own defence system.

Reef Fish

Coral reefs afford a number of different habitats, each of which is home to a characteristic set of fish species. Each type of fish exploits different aspects of their habitat so that a great many species can occupy the same ecological niche without undue competition.

Some, like the ubiquitous parrotfish, are herbivores and graze on the thin film of algae that grows on all bare surfaces or else nibble at the waving fronds of more leafy types of algae. Others, like the butterflyfish, dine on the coral itself. Puffer fish are able to digest even the calcareous exo-skeleton of the coral polyp and have no trouble with other variously-armoured invertebrates such as sea urchins, crustacea and starfish.

Most species of reef fish, however, are carnivores, ranging from the diminutive gobies who munch on minuscule crustacea, to the mightiest shark species which prey on large fish, turtles and even other sharks. Many of these killers are especially active at dawn and dusk when the crepuscular light beneath the waves makes them less visible to their potential prey.

Indian redfin butterflyfish
(*Chaetodon trifasciatus*)

Feather star

Cherry blossom

Pyjama cardinal
(*Sphaeramia
nematoptera*)

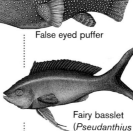

False eyed puffer

Fairy basslet
(*Pseudanthius
dispar*)

Crustaceans

Crustacea—crabs, lobsters and shrimps—occur in great abundance in coral reef habitats, though their presence is not always much in evidence on account of their tendency to hide away in caves and crevices, particularly by day.

Barracuda

The barracuda (*Sphyraena* spp.) is a rapacious predator on other fish. Most species occur in schools, with the exception of the great barracuda (*Sphyraena barracuda*) which hunts alone and can grow up to 190 centimetres in length and weighs some 40 kilograms. They seek their prey in a wide range of habitats from turbid inshore waters to the open sea.

The Ocean Depths

Many creatures of the open sea find their daily sustenance in the countless billions of tiny photosynthetic organisms known as phytoplankton that fill the world's oceans. Only the upper waters of the ocean receive sufficient light to enable the phytoplankton to go about their business and as one might expect, phytoplankton are especially abundant in tropical seas which receive plenty of sunlight—the waters off Bali are among the world's richest in this respect.

This invisible 'forest' of phytoplankton provides food for an equally numerous army of microscopic animals, the zooplankton, which feed on them. Together, they stand at the bottom of an oceanic food chain which feeds some of the mightiest creatures living today—the whales.

These behemoths of the ocean waves dine on plankton, squid, small fish and krill (tiny crustacea) which they filter from the sea by means of a giant sieve-like structure in their mouths, consisting of hundreds of slender whalebone plates, fringed with bristles, which are suspended from their upper jaws. Some 18 species of whale occur in Balinese waters, though they are unlikely to be seen from the shore. Dolphins, however, are quite a common sight, especially from boats.

Turtles

Marine turtles are quite at home in the deepest of waters but must come onto dry land to lay their eggs. Often, they will cross thousands of kilometres of ocean to return to their natal nesting beaches—it is thought that scent trails may assist them in this remarkable feat of navigation.

The seas around Bali are frequented by three species of turtle. They are the green turtle (*Chelonia mydas*),

Humpback whales

Female humpback whales (*Megaptera novaeangliae*) can reach lengths of up to 16 metres (males are slightly shorter). Found worldwide, they spend most of their time in the middle of the oceans, but migrate to shallow tropical waters for breeding–the gestation period is eleven and a half months. They are famous for their 'song' which consists of repeated sequences of sound which are arranged in complex patterns that can last for more than half an hour at a time. The song is sung by single males during the breeding season and varies between populations. Humpback whales are also well-known for 'breaching', that is, launching themselves out of the water only to fall back with a loud splash that can be heard a long way off. This habit may either be a form of display or some sort of communication. Their diet consists mainly of krill and small fish.

the leatherback (*Dermochelys coriacea*) and the hawksbill (*Eretmochelys imbricata*). They generally lay their eggs in sandy nests on the beaches of Bali's north-western coastline.

Hawksbill turtles also come ashore on the southern beaches of Nusa Dua, while green turtles visit the island of Nusa Penida. Marine turtles nest all year round, but seem to prefer to lay their eggs during the rainy season from September to December.

Green Turtle

The shell, or carapace, of the green turtle, like that of other members of the Chelonia family (tortoises and terrapins), comprises an outer layer of keratin and an inner bony structure fused to the spine and ribs. The soft underbelly is protected by the plastron, which again consists of fused ribs and bones derived from the shoulder girdle.

Squid

Squid are literally jet-propelled creatures who use a powerful combination of muscles working in opposition to each other to expel water from their body cavities, thereby effecting a forward motion. They are shaped like torpedoes for minimum drag, and lacking any inherent buoyancy, must keep swimming or else sink—position-maintenance speed is one tenth of the maximum possible.

Squid are also able to change their colour at will. Colour pigments, linked to muscle fibres, expand or contract in response to nerve signals, and when angry, alarmed or sexually aroused, many species pulse with vivid bands of colour of startling intensity, which sweep over the creature in iridescent waves.

Oceanic Whitetip Shark

The oceanic whitetip shark (*Carcharhinus longimanus*) cruises the surface layers of the open sea down to a depth of 500 metres. It grows to some 2 metres in length and is distinguished by its distinctive white-tipped fins. It is a ferocious predator and is extremely dangerous.

Dolphin

The common dolphin (*Delphinus delphis*) is distinguished by its tan or buff flanks. They are extremely fast swimmers, reaching speeds of up to 40 kilometres an hour. They feed mostly at night on deep-sea fish and squid but during the day they are very active, often leaping out of the water in play. The lifespan of the common dolphin is about 20 years and individuals often congregate in large populations of several thousand.

GLOSSARY

Annual	A flowering plant which completes its whole life cycle from seed germination to flowering and fruiting in a single growing season. This single season is generally assumed to take place in the course of one year though the actual developmental process may be far shorter in duration.
Bivalves	Any mollusk, such as a clam, mussel, oyster, or scallop, belonging to the class Bivalvia, possessing two shells hinged together, a soft body and gills.
Endemic	A species which has arisen evolutionarily in the place or region where it is currently found, rather than having been introduced there at some later moment in time.
Epiphyte	A plant growing above ground which is supported nonparasitically by another plant or object, and which derives its nutrients from the rain, leaf-fall, the activities of ants and other such processes.
Gamete	A mature reproductive cell, such as a sperm, which unites with its sexual counterpart (in this case an egg) to form a new organism.
Inflorescence	A cluster of flowers arranged on a single or branched stem.
Nymph	The young of an insect which has undergone incomplete metamorphosis.
Photosynthesis	The production of carbohydrates and other complex organic compounds from atmospheric carbon dioxide, water and inorganic salts, using sunlight as the source of energy.
Pioneer species	A plant or animal species which successfully establishes itself in a barren area, thereby initiating an ecosystem which is more conducive to supporting other life forms.
Plankton	Microscopic single-cell organisms which float in the sea (or lakes) and which obtain their energy from water-borne nutrients and by photosynthesis.
Polyp	A sedentary animal typically possessing a more or less fixed base, a columnar body and a free end equipped with mouth and tentacles.
Spadix	A spiky inflorescence (see above) which is typically enclosed by an outer sheath or spathe.
Substrate	The base or material on which a nonmotile organism lives or grows.
Transpiration	The means by which water passes through a plant, from the roots to the leaves, via the vascular system.
Understorey	The plants and shrubs which grow beneath the main canopy of a forest.
Zooplankton	Microscopic, non-photosynthetic animals, living in the seas and in lakes, which obtain their sustenance by feeding off plankton (see above).